Quake Chronicles: Navigating Earth's Powerful Tremors

White R.

Table of Contents

Chapter 1

Definition and Nature of Earthquakes

Welcome to the comprehensive exploration of seismic events and their significance in our world. In this chapter, we will delve into the definition, scientific explanation, the relationship between seismic activity and tectonic movements, and examine significant historical earthquakes and their profound impact on civilizations.

Definition and Scientific Explanation of Earthquakes

Earthquakes, the result of the sudden release of energy within the Earth's lithosphere, manifest as ground shaking or trembling. This seismic activity originates due to the movement of tectonic plates or volcanic activity, leading to the generation of seismic waves. These waves propagate through the Earth's surface, causing various levels of vibrations that can range from mild tremors to devastating quakes. Scientifically, earthquakes occur primarily due to the buildup and release of stress along fault lines, which are fractures in the Earth's crust. When the accumulated stress exceeds the strength of the rocks, it results in sudden movement, triggering an earthquake. The energy released during these events propagates through the Earth in the form of seismic waves.

Relationship between Seismic Activity and Tectonic Movements

Seismic activity is closely linked to the movement of Earth's tectonic plates, vast sections of the Earth's crust that float on the semi-fluid asthenosphere beneath them. The interactions between these plates at their boundaries, including convergent, divergent, and transform boundaries, are major contributors to seismic events. For instance, at convergent boundaries, where plates collide, intense pressure builds up, leading to frequent earthquakes.

Tectonic movements, such as subduction zones where one plate moves beneath another, or continental drift at transform boundaries, contribute significantly to seismic activities in these regions. These movements result in stress accumulation, subsequently releasing energy as earthquakes.

Historical Context: Significant Earthquakes and Their Impact
Throughout history, earthquakes have shaped civilizations and landscapes. Iconic seismic events, such as the 1906 San Francisco earthquake, the 2011 Tohoku earthquake and tsunami in Japan, or the devastating 2010 earthquake in Haiti, serve as reminders of their catastrophic impact.
These historical earthquakes, among many others, have left enduring imprints on societies, causing widespread destruction, loss of life, and socioeconomic disruptions. They've also spurred advancements in earthquake science, engineering, and disaster preparedness, contributing to our understanding of seismic risks and mitigation strategies.

Causes of Earthquakes

Earthquakes, among nature's most powerful and unpredictable phenomena, are primarily caused by several geological factors. Understanding these causes is pivotal in comprehending the mechanics behind seismic events.
The following are causes of earthquakes;

1. Plate Tectonics: Movementof Earth's Lithosphere and Boundaries

One of the primary causes of earthquakes lies in the movement of the Earth's lithospheric plates. The Earth's surface is divided into several large and small tectonic plates that float on the semi-fluid asthenosphere beneath them. These plates are in constant motion due to the intense heat generated by the Earth's core.

At plate boundaries, where these massive plates meet, various types of interactions occur. Subduction zones, where one plate sinks beneath another, create immense pressure and stress. When this stress exceeds the strength of the rocks along the boundary, it results in sudden movement and the release of seismic energy in the form of earthquakes. Convergent boundaries, where plates collide, and divergent boundaries, where plates move apart, also contribute significantly to seismic activity.

2. Fault Lines and Their Role in Earthquake Occurrences

Fault lines, fractures or zones of weakness in the Earth's crust, play a crucial role in the occurrence of earthquakes. These fractures occur due to the movement of tectonic plates and are characterized by their ability to accumulate stress. When the accumulated stress along these fault lines surpasses the rock's strength, it leads to a sudden release of energy, causing an earthquake.

The movement along fault lines can be vertical (normal faults), horizontal (strike-slip faults), or a combination of both. The San Andreas Fault in

California, for example, is a well-known strike-slip fault where the Pacific Plate and the North American Plate slide past each other, generating seismic activity.

Other Triggers Such as Volcanic Activity and Human-Induced Quakes
Aside from plate tectonics and fault movements, earthquakes can also be triggered by other factors. Volcanic activity, particularly during volcanic eruptions, can induce earthquakes. The movement of magma beneath the Earth's surface exerts pressure on the surrounding rocks, leading to tremors.

Chapter 2

The Science Behind
Earthquakes

Earthquakes, as manifestations of immense natural forces, stem from intricate geological processes that shape our planet. Chapter 2 embarks on an exploration of the underlying scientific principles driving seismic events, delving into the fascinating realm of seismic waves, measurement methods, and the forces that orchestrate these formidable natural phenomena.

Understanding the scientific principles governing earthquakes is pivotal in fortifying our preparedness and response to these natural upheavals. This chapter endeavors to unravel the intricacies of seismic science, fostering a deeper appreciation of the geological mechanisms driving earthquakes and their profound impacts on our world.

Seismic Waves and Their Effects

Seismic waves, the fundamental elements of earthquake propagation, play a pivotal role in understanding the behavior and impact of these natural phenomena.

Types of Seismic Waves

1. P-Waves (Primary Waves): These are the fastest seismic waves, characterized by their compressional motion, similar to sound waves. They travel through solids, liquids, and gases, causing them to expand and contract in the direction of wave propagation.

2. S-Waves (Secondary Waves): These waves travel slower than P-waves and move in a shearing or transverse motion. Unlike P-waves, S-waves can only travel through solids and are responsible for most of the structural damage during earthquakes.

3. Surface Waves: Surface waves, unlike P and S-waves, travel only along the Earth's surface. They are slower than both P and S-waves but have the most significant impact on buildings and structures.

Propagation of Seismic Waves and Their Impact on Structures

Seismic waves propagate through the Earth from the earthquake's epicenter, carrying energy that causes the ground to shake. As P-waves and S-waves pass through different materials, their speeds and ability to penetrate vary. P-waves move faster and can pass through both solid and liquid layers, while S-waves cannot travel through liquids, such as the Earth's outer core.

The effects of these waves on structures are profound. P-waves are less destructive but can cause subtle ground movements, alerting individuals seconds before the more damaging S-waves arrive. S-waves, being slower and transverse in nature, cause significant shaking and structural damage, especially to buildings not designed to withstand lateral forces. Surface waves, though slower, can cause the most prolonged shaking and result in the majority of structural damage during an earthquake due to their horizontal and vertical motion along the Earth's surface.

The Richter Scale and Other Measurement Methods for Earthquakes

The Richter Scale, developed by Charles F. Richter in 1935, quantifies the magnitude of earthquakes based on the amplitude of seismic waves recorded by seismographs. It's a logarithmic scale, meaning each whole number increase represents a tenfold increase in the amplitude of seismic waves and approximately 31.6 times more energy release.

Besides the Richter Scale, other scales like the moment magnitude scale (Mw) are used to measure earthquake magnitude, taking into account the seismic moment released during an earthquake. These scales aid in categorizing earthquakes, understanding their energy release, and assessing potential hazards.

Understanding these various seismic waves and measurement methods is crucial for evaluating earthquake strength, predicting potential damage, and implementing effective safety and structural measures.

Chapter 3

Impact and Consequences

The aftermath of seismic events such as earthquakes reverberates through communities, landscapes, and economies, leaving indelible marks on every aspect of life. Chapter 3 delves into the multifaceted impacts and consequences of earthquakes, shedding light on the physical, human, and environmental toll these natural phenomena exact.

The Profound Impacts of Earthquakes
Earthquakes unleash forces that transcend physical destruction, profoundly impacting societies,

ecosystems, and the well-being of individuals. This chapter unravels the far-reaching consequences, from structural damages to human casualties, and from environmental disruptions to the economic aftershocks that echo long after the ground ceases to shake.

The chapter navigates through the ripple effects of seismic events, unraveling the complex interplay of physical destruction, human suffering, environmental repercussions, and the socio-economic upheaval that often ensues. It underscores the imperative of comprehensive understanding and preparedness to mitigate these far-reaching impacts.

Physical and Structural Damages

Earthquakes unleash unforgiving forces that wreak havoc on buildings, infrastructure, and landscapes. The physical and structural damages, ranging from crumbling edifices to crumbling lifelines, serve as stark reminders of the sheer power seismic events possess to reshape the built environment.
 The following the physical and structural damages caused by earthquakes;

1. Overview of Damages to Buildings, Infrastructure, and Landscapes

Earthquakes unleash devastating forces that inflict severe damages on various structures and the environment. Buildings, bridges, roads, and critical infrastructure bear the brunt of seismic activity. They suffer from structural collapses, extensive cracking, and destruction, leading to massive economic losses and disruptions.

2. Structural Vulnerabilities and Collapse in Earthquake-Prone Regions

Regions situated along fault lines or tectonic plate boundaries face heightened risks. Structural vulnerabilities in these earthquake-prone areas exacerbate damages. Buildings and infrastructure often lack seismic-resistant designs, making them more susceptible to severe structural failure and collapse during seismic events.

3. Impact on Critical Facilities: Hospitals, Utilities, and Transportation Systems

The impact of earthquakes extends beyond structural damage. Critical facilities such as hospitals, utilities (water, electricity, gas), and transportation systems suffer severe disruptions. Damage to these vital services significantly impedes response and recovery efforts, amplifying the overall impact on affected communities.

Human and Environmental Impacts

Beyond bricks and mortar, earthquakes deeply affect human lives and the environment. The toll of injuries, fatalities, and displacement, juxtaposed against ecological upheavals like landslides and tsunamis, underscores the magnitude of devastation and the challenges communities face in the wake of seismic events.

Below are the impacts of earthquakes on humans and environment;

1. Effects on Human Lives: Injuries, Fatalities, and Displacement

Human lives bear the brunt of seismic disasters. Injuries, fatalities, and widespread displacement of populations occur due to collapsing buildings, falling debris, and the ensuing chaos during earthquakes. The immediate and long-term psychological and emotional impacts on survivors and affected communities are profound.

2. Environmental Consequences: Landslides, Tsunamis, and Ecosystem Disruptions

Earthquakes trigger secondary hazards such as landslides, avalanches, and tsunamis, particularly when the epicenter is near coastal areas. Landslides disrupt terrains, altering landscapes and threatening communities. Tsunamis, generated by undersea seismic activity, cause catastrophic flooding along coastlines. Additionally, seismic events disrupt ecosystems, affecting wildlife habitats and biodiversity.

3 Economic Ramifications and Societal Disruptions in Affected Regions

The economic toll of earthquakes is colossal. Societies experience disruptions in economic activities, trade, and productivity due to damages to industries, businesses, and infrastructure. The road to recovery is prolonged, impacting livelihoods, leading to economic recessions in affected regions, and necessitating substantial external aid and investments.

Case Studies and Historical Earthquakes

Studying historical earthquakes serves as a repository of invaluable lessons. Through notable seismic events of the past, we gain insights into the immediate aftermath, recovery efforts, and the long-term implications for societies and regions affected by seismic catastrophes.

1. Notable Historical Earthquakes and Their Aftermath

Studying historical earthquakes provides insights into their aftermath. Iconic seismic events like the 1906 San Francisco earthquake, the 2011 Tohoku earthquake, and the 2015 Nepal earthquake serve as stark reminders of their devastating impacts on societies, economies, and the environment.

2. Lessons Learned from Past Events: Recovery, Rebuilding, and Long-term Impact

Analyzing historical earthquakes aids in understanding the recovery and rebuilding efforts. It offers invaluable lessons in implementing robust disaster management plans, enhancing building codes, developing early warning systems, and fostering community resilience against future seismic events.

Chapter 4

Preparedness and Safety Measures

Preparedness and safety measures serve as the bedrock for minimizing risks and ensuring resilience in the face of seismic events like earthquakes. This chapter delves into a comprehensive array of strategies, guidelines, and practices designed to enhance readiness, protect lives, and reduce vulnerabilities during and after an earthquake.

The Essence of Preparedness
Preparedness encompasses a proactive approach aimed at equipping individuals, families, communities, and institutions with the knowledge, resources, and plans necessary to effectively respond to earthquakes. It involves understanding potential risks, establishing emergency protocols, and having the necessary tools and plans in place to mitigate the impact of seismic events.

Embracing Safety Measures
Safety measures form the cornerstone of earthquake preparedness. They encompass a wide spectrum of actions and practices geared towards minimizing risks, ensuring personal safety, and safeguarding structures and infrastructure against the destructive forces of earthquakes.

Importance of Preparedness and Safety Measures
In earthquake-prone regions, the importance of preparedness and safety measures cannot be overstated. Being prepared and implementing safety protocols can significantly mitigate the impact of earthquakes, reduce casualties, minimize property damage, and expedite recovery efforts.

Exploring Preparedness and Safety Measures
In this chapter, we explore various dimensions of preparedness and safety measures. From individual readiness plans and safety protocols to engineering

solutions, community-wide preparedness, and government initiatives, we delve into a diverse range of strategies essential for ensuring resilience and protecting lives and property during seismic events.

Personal Preparedness

1. Creating a Family Emergency Plan and Earthquake Survival Kit

Establishing a comprehensive family emergency plan is essential. This plan should include designated meeting points, communication strategies, and specific responsibilities for family members. Additionally, assembling an earthquake survival kit comprising essential items like non-perishable food, water, first-aid supplies, flashlights, and important documents ensures readiness during emergencies.

2. Safety Measures During an Earthquake: Drop, Cover, and Hold On

Educating individuals on safety measures during earthquakes is crucial. The "Drop, Cover, and Hold On" technique instructs people to drop to the ground, seek cover under sturdy furniture or against interior walls, and hold on until the shaking stops, reducing the risk of injury from falling objects or debris.

3. Knowing Evacuation Routes and Emergency Contacts

Familiarity with evacuation routes and emergency contacts is pivotal. Identifying safe evacuation paths and having readily accessible emergency contacts ensures swift and coordinated responses during and after seismic events.

Engineering and Building Safety

1. Earthquake-Resistant Construction Techniques
Implementing earthquake-resistant construction techniques is crucial for mitigating structural damages. Employing seismic isolation, base isolation, reinforced concrete structures, and flexible building materials strengthens structures to withstand seismic forces, reducing the likelihood of collapse or damage.

2. Retrofitting Existing Buildings for Increased Seismic Safety

Retrofitting existing buildings involves reinforcing structures to enhance their resilience against earthquakes. Installing shear walls, bracing systems, and using damping devices in older buildings significantly improves their ability to withstand seismic forces.

3. Role of Architectural Design in Reducing Earthquake Damage

Architectural design plays a pivotal role in minimizing earthquake damage. Incorporating flexible and resilient designs, like open floor plans, in structures allows for better absorption of seismic forces, reducing the risk of structural failure.

Community Preparedness and Planning

1. Community-Wide Emergency Plans and Drills
Developing community-wide emergency plans and conducting regular drills fosters preparedness among residents. These plans should outline evacuation procedures, assembly points, and communication

channels, ensuring coordinated responses during seismic events.

2. Evacuation Procedures and Designated Safe Zones

Establishing clear evacuation procedures and designating safe zones is critical for community safety. Identifying safe assembly areas away from potential hazards and communicating evacuation routes ensures organized and efficient evacuations.

3. Coordination Among Local Authorities, Responders, and Citizens

Collaboration among local authorities, emergency responders, and citizens is vital. Establishing communication channels, holding community meetings, and fostering partnerships enable effective coordination and response efforts during earthquakes.

Chapter 5

Mitigation and Risk Reduction

Mitigation and risk reduction stand as the cornerstone in minimizing the devastating impacts of seismic events like earthquakes. These concepts encompass a diverse range of strategies and measures aimed at lessening the vulnerability of communities, infrastructure, and ecosystems to seismic hazards.

What is Mitigation?
Mitigation refers to a proactive approach that focuses on lessening the severity and potential consequences of natural disasters, such as earthquakes. In the context of earthquakes, mitigation strategies aim to reduce the impact of seismic events on people, buildings, infrastructure, and the environment. It involves deploying various measures to enhance resilience, strengthen structures, and implement policies that mitigate the potential damage caused by earthquakes.

Understanding Risk Reduction
Risk reduction involves a comprehensive set of actions and policies designed to decrease the probability of an earthquake causing significant damage, loss of life, and disruption. It encompasses efforts to minimize vulnerabilities and exposures to seismic hazards, thus reducing the overall risk posed by earthquakes. This involves integrating scientific knowledge, community preparedness, infrastructure improvements, and effective governance to mitigate the potential impact of seismic events.

Why Mitigation and Risk Reduction Matter

The significance of mitigation and risk reduction in the context of earthquakes cannot be overstated. By implementing these strategies, communities and nations can significantly reduce the socio-economic impacts of seismic events. This involves not only building structures that can better withstand earthquakes but also developing policies, early warning systems, and preparedness measures to ensure swift responses and effective recovery.

Exploring Mitigation and Risk Reduction
In this chapter, we delve into various facets of mitigation and risk reduction strategies. We examine innovative engineering solutions, governmental policies, urban planning initiatives, international collaborations, and technological advancements aimed at enhancing preparedness, reducing vulnerabilities, and mitigating the impacts of earthquakes. Understanding these critical aspects is fundamental in developing resilient communities capable of withstanding the challenges posed by seismic events.

Engineering Solutions for Mitigation

1. Innovative Technologies and Materials for Seismic Resistance

Innovative advancements in construction technologies and materials contribute significantly to seismic resistance. Technologies like base isolation systems, seismic dampers, and advanced materials (such as carbon fiber composites and flexible steel frames) enhance a structure's ability to withstand seismic forces.

2. Retrofitting Infrastructure and Critical Facilities for Resilience

Retrofitting existing infrastructure and critical facilities is essential for increasing resilience. This involves applying seismic upgrades, reinforcing structural elements, and implementing modern engineering techniques to enhance their ability to withstand earthquakes.

3. Challenges and Successes in Implementing Earthquake-Resistant Designs

Implementing earthquake-resistant designs presents challenges, including cost considerations, technical complexities, and retrofitting older structures. Highlighting success stories in deploying such designs provides insights into overcoming obstacles and encouraging broader adoption.

Government Policies and Urban Planning

1. Land Use Planning to Mitigate Seismic Risks

Effective land use planning plays a pivotal role in mitigating seismic risks. Zoning regulations, land-use policies, and avoiding construction in high-risk areas contribute to reducing vulnerability to earthquakes.

2. Building Codes, Zoning Regulations, and Safety Standards

Stringent building codes and safety standards are crucial. Governments enforcing rigorous codes for construction, retrofitting guidelines, and zoning regulations ensure that structures meet specific seismic safety requirements.

3. Strategies for Retrofitting High-Risk Areas and Vulnerable Structures

Strategies aimed at retrofitting high-risk areas and vulnerable structures involve prioritizing areas most susceptible to earthquakes. Implementing phased retrofitting programs and incentivizing compliance encourage the enhancement of structural resilience.

International Efforts and Early Warning Systems

1. Collaborative Initiatives for Global Earthquake Monitoring

International collaborations in earthquake monitoring and data sharing facilitate a comprehensive understanding of seismic activities worldwide. Collaborative efforts among various countries enhance the accuracy of earthquake forecasting and risk assessments.

2. Development and Implementation of Early Warning Systems

Investments in early warning systems are crucial for reducing earthquake impacts. Development and deployment of real-time warning systems, seismic monitoring networks, and public alert dissemination contribute to enhancing preparedness.

3. How Technology Aids in Predicting Earthquakes and Reducing Risks

Technological advancements aid in predicting earthquakes and reducing associated risks. Innovative approaches, including seismological research, satellite monitoring, and machine learning algorithms, significantly contribute to early detection and risk reduction.

Chapter 6

Earthquakes and the Future

This chapter sets its gaze on the future of seismic research, forecasting the trajectory of advancements, challenges, and transformative efforts to fortify global resilience against seismic events. This chapter navigates through the evolving landscape of earthquake studies, showcasing the innovations, education initiatives, and collaborative endeavors shaping our preparedness for the seismic challenges ahead.

Advancements in Earthquake Research

1. Innovations in Earthquake Prediction and Forecasting

The future of earthquake research unfolds with promising innovations in prediction and forecasting techniques. From harnessing artificial intelligence and machine learning algorithms to augmenting seismic monitoring networks, these advancements hold the promise of refining our ability to forecast seismic events with greater accuracy and lead time.

2. Technology's Role in Early Detection and Warning Systems

Technology emerges as a beacon in early detection and warning systems. Expanding seismic networks, real-time data analysis, and the fusion of remote sensing technologies offer prospects for more robust early warning systems. These technological strides pave the way for timely alerts, equipping communities with crucial seconds to enact life-saving measures.

3. Future Prospects and Challenges in Seismic Studies

As seismic studies progress, the horizon presents a tapestry of prospects and challenges. While technological leaps propel our understanding and preparedness, challenges persist in decoding precursory signals, refining models, and fostering trust in forecasting. Navigating these challenges shapes the course of future seismic endeavors.

Sustainable Approaches and Education

1. Education and Awareness Campaigns for Earthquake Preparedness

Education emerges as a linchpin in cultivating resilient communities. Extensive awareness campaigns, comprehensive education, and knowledge dissemination empower individuals to embrace preparedness. Equipping communities with information fosters a culture of resilience, laying the foundation for safer responses to seismic events.

2. Sustainable Approaches to Minimize Seismic Risks

Incorporating sustainable practices stands as a cornerstone in mitigating seismic risks. Integrating resilient designs, enforcing stringent building codes, and adopting sustainable land use policies play pivotal roles. These practices fortify structures and communities, minimizing vulnerabilities against seismic hazards.

3. Collaborative Efforts for Global Resilience Against Earthquakes

Global resilience hinges on collaborative initiatives transcending borders. Partnerships, knowledge-sharing networks, and collaborative platforms foster international cooperation. Through shared experiences, best practices, and expertise exchange, these endeavors fortify global preparedness against seismic events.

Conclusion

As we draw the curtains on this comprehensive exploration of earthquakes, it becomes evident that these seismic events, though formidable, don't solely define our fate. Rather, our understanding, preparedness, and collective efforts shape our resilience in the face of Earth's tremors.

Throughout this guide, we've navigated the seismic landscape, uncovering the intricate science behind these natural phenomena. From the fundamental mechanisms driving earthquakes to the profound impacts rippling across societies and landscapes, each chapter has unveiled the multifaceted nature of seismic events.

But beyond mere understanding lies the cornerstone of resilience — a tapestry woven with knowledge, preparation, innovation, and collaboration. The future holds promising innovations in prediction, technology's crucial role in early warning systems, and sustainable approaches that fortify communities and infrastructure.

Education emerges as a beacon, empowering individuals with the knowledge to embrace preparedness. Sustainable practices and stringent building codes stand as bastions, fortifying structures and communities against seismic upheavals.

Yet, resilience isn't forged in isolation. It thrives in collaboration, transcending boundaries and fostering global solidarity. Collaborative efforts, partnerships, and knowledge exchange networks pave the way for a more resilient world against Earth's tremors.

As we close this chapter, let us not forget the lessons learned from history's seismic moments, the strides made in research, and the resilience showcased by communities facing the earth's relentless movements. Let us continue to innovate, educate, collaborate, and prepare, ensuring that our future generations thrive in a world better equipped to navigate the tremors that define our planet.

In this ongoing journey towards resilience, may our understanding and preparedness be the bedrock that steadies us, guiding us towards a safer, more resilient future in the face of Earth's powerful tremors.